Who You Are
and How You Came to Be

**Other books by
Stephen Hawley Martin**

*LIFE AFTER DEATH,
POWERFUL EVIDENCE YOU
WILL NEVER DIE*

*LIFE AFTER DEATH BOOK II,
Heaven, Hell, & You*

*THE MT PELEE REDEMPTION
A Thriller*

*DEATH IN ADVERTISING
A Whodunit*

*THE SEARCH FOR NINA FLETCHER
Romantic Suspense*

*THE COLOR OF DEMONS
A Paranormal Thriller*

*AMAZING TRUTH
AND HOW TO PROFIT FROM IT*

*A WITCH IN THE FAMILY
An Award-Winning Author Investigates
His Ancestor's Trial and Execution*

*REINCARNATION
Good News for Open-Minded Christians
& Other Truth-Seekers*

*THE PROSPERITY FACTOR,
SECRET OF THE SAGES*

*YOU CAN BE HAPPY, NO JOKE
Self-Help / Happiness*

*HOW TO MASTER LIFE
Self-Help / Body, Mind & Spirit*

Who You Are
and How You Came to Be

by

Stephen Hawley Martin

WWW.OAKLEAPRESS.COM

Who You Are and How You Came to Be © 2017 by Stephen Hawley Martin. All rights reserved. No part of this book may be used or reproduced in any manner whatsoever without written permission except in the case of brief quotations embodied in critical articles and reviews. For information address The Oaklea Press Inc., 41 Old Mill Road, Richmond, Virginia 23226.

CONTENTS

FOREWORD.. 9

Chapter One
A Theory of Human Origins... 9

Chapter Two
Supporting Facts about Consciousness........................... 22

Chapter Three
My Theory about the Origin of Physical Reality............. 27

Chapter Four
What to Do about What You Now Know....................... 36

Chapter Five
How to Find Your Purpose... 44

Chapter Six
The Return of a Very Old Worldview............................. 57

"Whatever you can do or dream you can, begin it. Boldness has genius, power, and magic in it."

— Johann Wolfgang von Goethe (1749-1832)

FOREWORD

It has been said that good things come in small packages, which is my intension for this book. At 15,500 words, it is indeed a small package, given that most books range from 40,000 up to more than 100,000 words. Nevertheless, I believe this book can impart knowledge and wisdom you can use to make your life as fulfilling and satisfying as it can be.

Once you know who you are and how you came to be, you will be equipped to self-actualize fully within the practical limits you have to deal with. As a sage once said, no matter where you want to go, you have to start where you are now.

Here are the steps I suggest you take:

1. Read this book. Once you have finished, you may benefit from reading it a second time.

2. Determine the limitations you will have to deal with such as your familial obligations and your financial resources.

3. Determine your Dharma, and write it down. (You will learn about Dharma in Chapters Four and Five.)

4. Create a vision of where you want to go and what you want to accomplish. Write it down.

5. Create a step-by-step plan and timetable to take you there.

6. Review your vision, your plan, and your timetable at least weekly, if not daily, and measure your progress. If a door closes, look for a door that is open. Revise your plan accordingly, and never give up.

You may only be able to practice your Dharma on a part-time basis now and for some period of time going forward. Nevertheless, it is important to begin using it, honing it, and working toward a time when exercising it can become your full-time occupation.

Don't put this off. It may be a cliché, but it's true: Life is short.

Now is the time to begin.

Stephen Hawley Martin

Chapter One
A Theory of Human Origins

Based on indisputable evidence, we now know the brain does not create consciousness. Rather, the brain is a wireless receiver that captures consciousness and integrates it with the body.

Having spend a good deal of time considering the implications of this finding by researchers at the Division of Perceptual Studies at the University of Virginia School of Medicine, it now seems clear to me how you and I came to be. Before going into it, however, let's consider why scientists did not understand this before now.

More than 300 years ago modern science inadvertently got boxed in by an English philosopher named Thomas Hobbes (1588-1679) when Hobbes argued that physical reality—what can be seen and measured—is all that exists. "The universe," Hobbes wrote, "that is, the whole mass of things that are, is corporeal, that is to say, body, and hath the dimensions of magnitude, namely length, breadth, and depth; also every part of body is likewise body, and hath the like dimensions, and consequently every part of the universe is body, and that which is not body is no part of the universe: and because the universe is all, that which is no part of it is nothing, and consequently nowhere."

We now know, of course, many things Hobbes and others of his day did not. A great deal exists that cannot be seen or easily measured, things that do not fall into the category of "body," such as radio waves, gamma rays, gravity, atoms, quarks, electrons, and other subatomic particles. But Hobbes' musing made sense at the time and eventually became scientific dogma that is now, figuratively speaking, engraved in granite. This has created a box many scientists dare not think outside. In the past, whenever a brave soul did think outside the box, he or she was shouted down or ridiculed by true believers. Considering what researchers at the University of Virginia have determined, however, I'm hopeful such zealots will now think twice before making fools of themselves. As British bio chemist Rupert Sheldrake is said to have quipped, "[Today's Scientific Materialists say] give us one free miracle and we'll explain the rest.' And the one free miracle is the appearance of all the matter and energy of the universe, and all the laws that govern it, from nothing in a single instant."

In defense of Thomas Hobbes, it should be noted that the doctrine of scientific materialism he unleashed on the world has resulted in a great deal of good. Hobbes lived at a time when superstitions, magical thinking, and belief in witchcraft were rampant. Hobbes was railing against beliefs that caused a great deal of misery and destruction. I am personally aware of this because my ancestor, Susannah North Martin (1621-1692), was hanged for witchcraft in on July 19, 1692 in Salem, Massachusetts, an event that cast a shadow over the Martin family for decades and caused me to doubt the veracity of expert opinions from an early age. After all, it was the experts of the day such as Cotton Mather and John Hathorne, whose descendent Nathanial Hawthorne added a W to the name to disassociate himself from the Puritan judge, that sent Susannah and others to the gallows.

It's no wonder thinkers of Hobbes' day latched onto Hobbes' idea—so much so, in fact, that it eventually led to the Age of Reason, to the end of witch trials, to the scientific method, and arguably to the many good things and labor-saving devices that have come about because of modern science. But despite the positive results, modern science is based on a misconception that ultimately led to the erroneous belief that the brain creates consciousness even though plenty of evidence existed before UVA went public with its findings that life forms without brains are conscious to a degree. Earthworms, for example, have no brain, but are conscious enough to recoil when poked. Amoebae seem to know when food is passing by because they eat it. Plants grow toward the sun so they must be conscious of light or heat. Yet scientific dogma maintains that consciousness did not exist until evolution produced a brain, which accepted science continues to maintain came about as the result of an incredible series of accidents and what Charles Darwin dubbed, "Natural Selection."

No doubt Natural Selection is one mechanism at work in evolution, but surely it is not the only one. At the very least, something is missing from the theory. It seems to me illogical to think that even though the universe operates by complex laws, intelligence did not exist until evolution produced the extraordinarily complex brain found in humans. I doubt anyone who has deeply considered this assertion actually believes it. More than forty years ago, for example, I read a book that opened my eyes to the truth. Published in 1975, it refuted the idea that intelligence, consciousness, and awareness, came about as a result of evolution. Rather, it presented a strong case for turning that hypothesis on its head. Entitled *Intelligence Came First*, it was compiled and edited by Ernest Lester Smith, a Fellow of the Royal Society—the prestigious scientific academy of the United Kingdom dedicated to promoting excellence in science—and it caused quite a lot of controversy in its day.

The premise of the book is that, throughout eons of evolution needs have preceded the organs through which they are fulfilled—eyes, ears, taste buds, hearts, kidneys, livers, and so forth. Since each new organ developed in response to a need, why would the brain be an exception? Smith and his colleagues put forth a compelling argument that intelligence is primordial, quite able to function in its own realm. We now know this to be true thanks to the Division of Perceptual Studies at the University of Virginia School of Medicine. When one accepts this, all sorts of things make sense that modern science continues to ignore by labeling them anomalies, including ESP, psycho kinesis, and other paranormal phenomena.

How Humans Came to Be

Here is what resonates with me. Behind and forming reality is a creative force, the opposite of entropy, that is in fact consciousness itself.

For anyone who may not be aware, entropy, also known as the Second Law of Thermodynamics, refers to the idea that everything in the universe eventually moves from order to disorder. In other words, entropy is the invisible force that causes things to deteriorate. It's the reason your twenty-year-old car will eventually wear out, break down, and if you aren't careful, may leave you stranded somewhere. Your credit card and the mechanic you pay with it to fix your old car can be compared to the opposite of entropy—the force I'm talking about. I consider it to be the "Life Force," a term I have yet to find in any college textbook perhaps because, as stated by Wikipedia, "Life Force" is a concept in spirituality and alternative medicine. Ignoring it, however, does not make it go away. Life Force is what causes a wound to heal and a tiny acorn to grow into a mighty oak. It's what animates a mouse and an elephant. It causes seeds to germinate and is constantly pushing young animals to grow. You came from it. I came from it. Everything alive came from it, and everything—you and I included—is still connected to and part of it. That Life Force is conscious can be seen clearly in a field of sunflowers as they follow the sun from dawn to dusk across a summer sky.

It appears that everything alive—everything animated by Life Force—must constantly evolve or die. As the opposite of entropy, it is the force behind evolution. Think about it. As the case made in Ernest Lester Smith's 1975 book states, throughout the eons of evolution of life on earth, needs have preceded the organs through which they were fulfilled. Your high school biology teacher may push back on that, but it certainly makes sense. Even Charles Darwin doubted natural selection was the only mechanism at work in evolution. He said, for example, that to think the eye had evolved by natural selection "seems, I freely confess, absurd in the highest possible degree."

Who You Are and How You Came to Be

The point I'm leading to is this: Now that researchers at UVA have concluded after fifty years of meticulous research that the brain does not create consciousness, but rather connects consciousness to and integrates it with the body, it logically follows that brains evolved in order to capture consciousness and to bring it into physical reality.

Okay, you say, so what?

This provides the basis for my answer to the question, "Who are you and how did you come to be?" Here it is:

You are an individuated unit of consciousness that has come to believe you are separate from the whole.

Although conscious, primitive life forms—certainly one-celled animals, but even higher ones such as a sunflower or an earthworm—do not think or feel they are separate from the Force. They possess subjective-mind awareness only. In other words, they do not, and cannot, think about themselves.

At a point in evolution, however, we humans developed objective-mind awareness. We were able step outside ourselves, ignore our instincts, ponder our own existence, and make decisions for ourselves, decisions that were contrary to what the "still, small voice within" may have been telling us. This first happened at the time in human evolution recounted in the allegory found in the Bible's Book of Genesis, when one of the first humans, Eve, exercised free will perhaps for the first time by eating the fruit from the tree of the knowledge of good and evil.

Few would argue that there are not two kinds of thought. We might call them lower and higher, or subjective and objective because what differentiates the higher from the lower is the recognition of self. The plant, the worm, and probably a goldfish possess the lower kind only. They are unaware of self. Perhaps higher forms of mammals—chimpanzees, maybe dogs, and certainly human beings—possess both. You might say the higher variety of self-aware thought is possessed in progressively larger amounts as if ascending a scale.

The lower mode of thought, the subjective, is the intelligence or mind present everywhere underlying physical reality that, among other things, supports and controls the mechanics of life in every species and in every individual. It causes plants to grow and to push their roots into the soil. It causes hearts to beat and lungs to take in air. It controls all of the so-called involuntary functions of the body.

That this lower form of thought is everywhere at once coincides with Carl Jung's theory, which maintains that we humans share a universal subconscious mind. Moreover, we each have our own portion, our individual subconscious mind that blends into the collective, Infinite Mind. We have a conscious mind that makes us self-aware, and a subconscious mind that contains everything we have experienced in this life and perhaps, as we will soon discuss, past lives. The two types of mind are inextricably linked in that our conscious mind and personality of this lifetime arises out of the subconscious, which in turn arose out of the universal subjective mind. The gradual emergence of self-aware thought out of the universal subjective mind is implicit in the evolution of life from one-celled animals to increasingly higher forms, all the way to the current pinnacle, the minds of Homo sapiens.

How You Came to Be

Now you know who you are. You are a unit of individuated consciousness and though you may not realize it, you are still connected to and part of the whole. As mystics have been saying since the dawn of recorded history, "All is One."

In addition, while I suppose it may be possible, it seems unlikely to me that at the time of your birth your brain grabbed some subjective-mind consciousness out of the Life Force in order to form the human being you are today. Rather, it seems much more likely that this is the most recent of a long series of sojourns you have taken on this planet, perhaps stretching back to the beginnings of life on Earth. This possibility is supported by additional findings by UVA researchers. Since the 1960s, they have compiled more than 2500 cases of children who accurately remembered past lives. I have written a book that contains a good deal of information concerning the science of reincarnation, including an exploration of evolving morphogenetic fields and how they interact with genes, and so I will not go into detail about it at this time. I will, however, relate a bit of anecdotal evidence that may help you get your mind around how you came to be the unique individual you are today.

If you are old enough, you may remember Glenn Ford (1916-2006), a movie actor who flourished during Hollywood's Golden Age. Once, when he was approached about taking a role in a movie about Dutch psychic Peter Hurkos (1911-1988), Ford decided he ought to learn something about the paranormal. So, at the age of 54, Ford personally viewed demonstrations by Hurkos, conducted a number of interviews with experts on the subject, and in December 1975, voluntarily underwent three past-life regression sessions via hypnosis, during which he described five previous lives. As is typically done in such sessions, the hypnotized actor was regressed back to childhood, and then was coaxed further back before his current birth to recall previous lives.

Who You Are and How You Came to Be

In one session, Ford described himself as a bachelor music teacher named Charles Stewart of Elgin, Scotland who died in 1892. Stewart loved horses but hated his job teaching music to young schoolgirls. Amazingly, under hypnosis, Ford agreed to demonstrate his musical skill, and played passages from Beethoven, Mozart, and Bach. When Ford listened to the tapes of the interview, he said he shared Stewart's love for horses and had, since his early years, been considered a natural with those animals. Most significantly, however, he said that in his current life he did not, and could not, play the piano. Following the past-life regression sessions, researchers went to Scotland and located historical records of a music teacher named Charles Stewart of Elgin, Scotland who died in 1892.

A second hypnotic regression session with Ford brought out memories of a life as a member of French King Louis XIV's elite horse cavalry. Under hypnosis, Ford not only gave accurate information about his surroundings in France 300 or so years prior, he was able to speak French fluently—although in his current incarnation, he did not know the language.

Subsequently, Ford was regressed to other previous lives, describing a Christian martyr killed by lions in the Coliseum in third century Rome, and a seventeenth Century Royal Navy sailor who died of the Great Plague. In

Who You Are and How You Came to Be

his most recent prior lifetime, Ford was a cowboy who herded cattle in the American West.

It is interesting to note that although Ford starred in 106 movies, as well as several TV series ranging from comedy to police dramas to war stories, he was best known for and most often cast as a cowboy in Westerns.

What does this suggest about you? In the more than 200,000 years or so since the first homo sapiens walked planet earth, you have probably lived dozens of times, perhaps hundreds, and maybe even thousands. If that is indeed the case, and it seems likely to me, you have been evolving, sometimes rapidly, and in some lifetimes, perhaps not so much. In some lives you have been a man and in others, a woman. If you don't feel comfortable with the sex you came into this life with, perhaps in your most previous past life, you were a member of the opposite sex. If you have unexplained phobias, fears, or predilections—even talents as Glenn Ford did with horses—they may stem from events or conditioning that took place in a previous life. Research by UVA indicates this sort of thing is commonplace. Whatever the case may be, everything you have experienced is now part of you in what some would say is your subconscious mind, and what others would call your Soul.

Let me add, the key to happiness and to a truly successful life is for your ego self of this lifetime to join forces with your higher self in a spirit of cooperation. More will be written about this in an upcoming chapter. I'll finish this one by saying, we are and remain fully integrated components of universal consciousness. As such, our individual consciousness is eternal and our individual potential is that of the entire Infinite Mind. To what extent we achieve that potential through self-improvement and self-actualization is up to us.

Let me put that a different way. Individuated units of consciousness like you, once created, are immortal and have an unlimited capacity to evolve, each by developing and following his or her own path. As Jesus is quoted as having said in John 10:34, "Is it not written in your Law, 'I have said you are "gods?"'" (NIV translation; Jesus was referencing Psalm 82:6)

That's pretty heavy stuff, but I think it's true.

Chapter Two
Supporting Facts about Consciousness

I give a fairly in-depth summary of UVA's Division of Perceptual Studies findings about consciousness and the brain in the second edition of my book, *Life After Death, Powerful Evidence You Will Never Die,* and so in this book I will give only a brief overview. If you are so inclined, you can find a great deal more information on YouTube. A number of videos have been posted about the research at UVA, including panel discussions and on-camera interviews with Bruce Greyson, M.D., the Chester Carlson Professor of Psychiatry and Director of the Division of Perceptual Studies at the University.

As previously discussed, the bottom line takeaway is that brains do not actually create consciousness, despite what many scientists may still think. Dr. Greyson does say, however, that this mistaken belief is understandable since to an observer it does appear that the brain produces consciousness. Consider what happens when a person drinks too much or gets knocked on the head. Also, it's possible to measure electrical activity in the brain during certain kinds of mental tasks and to identify correlations between different areas of the brain and the different activities. We can stimulate different parts of the brain and record what experiences result, and we can remove parts of the brain and observe the results on behavior. This suggests that the brain is involved with thinking, perception, and memory, but according to Dr. Greyson, it does not mean the brain causes those thoughts, perceptions, and memories. What the measurements actually show are correlations, rather than causation. The truth is that thoughts, perceptions, and memories, actually occur somewhere else and then are received and processed by the brain in a way similar to how a television, cell phone, or radio receiver works.

Western science, Dr. Greyson pointed out in one of the interviews, is largely reductionist. It breaks everything down to its component parts, which are much easier to study than the whole, but the component parts do not always act like the whole. The brain is composed of millions of nerve cells or neurons, but a single neuron cannot formulate a thought, cannot feel angry or cold. It appears that brains can think and feel, but brain cells cannot. No one knows how many neurons are needed in order for them to collectively formulate a thought, nor do we know how a collection of neurons can think when a single neuron cannot.

Scientists get around this by saying consciousness is an emergent property of brains, a property that emerges when a large enough mass of brain cells gets together. According to Dr. Greyson, however, saying something is an emergent property is a way of saying it is a mystery that cannot be explained. It is a fact that there is no known mechanism in the brain or anywhere else that can produce non-physical things like thoughts, memories, or perceptions. The materialistic understanding of the world fails to deal with how electrical impulses, or a chemical trigger in the brain, can produce a thought or a feeling, or for that matter, anything the mind does. Despite this, according to Dr. Greyson, most scientists continue to maintain what he labeled, "The nineteenth century, materialist view that the brain in some miraculous way we do not understand produces consciousness." These scientists, he said, "Discount or ignore that consciousness in extreme circumstances can function very well without a brain."

Dr. Greyson noted that the idea the mind and the brain are separate is what most people believed until a couple of hundred years ago, but in the nineteenth century western world, beginning with the Darwinians, science began exploring the idea that the physical brain might be the source of thoughts and consciousness. Ironically, as one group of scientists attempted to explain consciousness in terms of Newtonian physics, scientists in a different discipline, physics, were forced to move away from Newtonian physics and develop quantum mechanics in order to explain phenomena in which consciousness—what a researcher knows or doesn't know—completely changes the results of certain experiments. It is as though the right hand did not know what the left hand was up to. Incredibly, this remains how things are today.

In one interview that can be found on YouTube, Dr. Greyson lists a number of examples of evidence researchers with the Division of Perceptual Studies have collected that demonstrate that consciousness can exist without a brain's involvement. It falls into four categories:

1. Recovery of lost consciousness in the moments or days prior to death among people who have been unconscious for prolonged periods of time.
2. Complex consciousness ability in some people who have minimal brain tissue.
3. Complex consciousness in near-death experiences when the brain is not functioning or is functioning at a greatly diminished level.
4. Memories, particularly among young children, accurately recalling details of a past life.

The first three categories above are discussed in detail in the book I mentioned at the beginning of this chapter. Also covered in that book as well as in my book on reincarnation are details concerning Number Four above, the Division's research into children's memories of past lives. Researchers at the University of Virginia have been conducting these investigations for more than fifty years, beginning with those by Dr. Ian Stevenson (1918-2007), and as a result have in excess of 2500 cases in their files as stated earlier. I have twice interviewed one of the Division's key researchers who has written two books on the reincarnation research findings, Jim B. Tucker, M.D., a child psychiatrist.

Anyone with an open mind who looks into what has been found will find it difficult to refute that reincarnation can and does happen. Perhaps the most convincing case study is that of James Leininger who recalled in vivid detail his life and death as a U.S. Army Air Force fighter pilot, James Huston, who was shot down in 1945 during the battle of Iwo Jima. As a small child, young James not only recalled a host of details of that life, including the name of his aircraft carrier and the type of plane he flew, he recalled the names of many of his shipmates and was able to recognize them and address them by name at a reunion of the aircraft carrier's crew.

In the next chapter I will go out on a limb to state another theory of mine that flies in the face of what today is accepted science.

Chapter Three
My Theory about the Origin of Physical Reality

In Chapter One, I stated my theory of human origins. In other works of mine, I have also stated another theory that flies in the face of materialistic science—my theory that the Universal Mind actually creates physical reality. A physics experiment called the "Double Slit Experiment" and the strange but revealing phenomenon associated with it known as "The Participating Observer," opened my eyes to this. The implications of the results of this experiment are straightforward and make perfect sense in light of what UVA has revealed, and yet scientists have failed to grasp them due the corner they find themselves boxed into, which was created by Thomas Hobbes.

Scientists have known for more than a hundred years that light can behave both as waves and as particles (photons), but until 1905 they thought it was comprised only of waves. Thomas Young (1773-1829) demonstrated in 1803 that light is waves by placing a screen with two parallel slits between a source of light—sunlight coming through a hole in a screen—and a wall. Each slit could be covered with a piece of cloth. These slits were razor thin, not as wide as the wavelength of the light. When waves of any kind pass through an opening not as wide as they are, the waves diffract. This was the case with one slit open. A fuzzy circle of light appeared on the wall.

When both slits were uncovered, alternating bands of light and darkness appeared, the center band being the brightest. Scientists call this a zebra pattern. The areas of light and dark resulted from what is known in wave mechanics as interference. Waves overlap and reinforce each other in some places and in others they cancel each other out. The bands of light on the wall indicated where one wave crest overlapped another crest. The dark areas showed where a crest and a trough met and canceled each other out.

In 1905, Albert Einstein published a paper that revealed light also behaves as if it consists of particles. He did so by using the photoelectric effect. When light hits the surface of a metal, it jars electrons loose from the atoms in the metal and sends them flying off as though struck by tiny billiard balls. So, Thomas Young demonstrated light is waves, and Einstein demonstrated it is particles. This sort of paradox is what led scientists to develop quantum mechanics.

Now let's consider a double slit experiment constructed to determine what happens when those conducting the experiment observe or do not observe which slits the photons of light pass through. This time a gun is used that fires one photon at a time. I first read about this experiment more than twenty years ago in an article entitled, "Faster Than What?" in the June 19, 1995 issue of *Newsweek* magazine. It reported on a paper to be published by a well-known quantum physicist, Raymond Chiao, then teaching at the University of California at Berkeley.

Much later, in July 2008, I reviewed the facts of this experiment as they are about to be presented here with a guest on a radio show I hosted at the time, the noted quantum physicist Henry P. Stapp, author of, *MINDFUL UNIVERSE: Quantum Mechanics and the Participating Observer* (Springer, 2007). He said I understood them correctly.

Both slits were open and a detector was used to determine which slit a photon passed through. A record was made of where each one hit. Only one photon at a time was shot, so one would suppose there could be no interference. This was the case. The photons did not make a zebra pattern. Rather, they made marks, tiny dots, on a screen.

When the detector was turned off, however, and it was not known which slit a photon passed through, the zebra pattern appeared. In other words, without the detector the particles behaved like waves even though they were fired one at a time. Imagine the stir this caused among

Who You Are and How You Came to Be

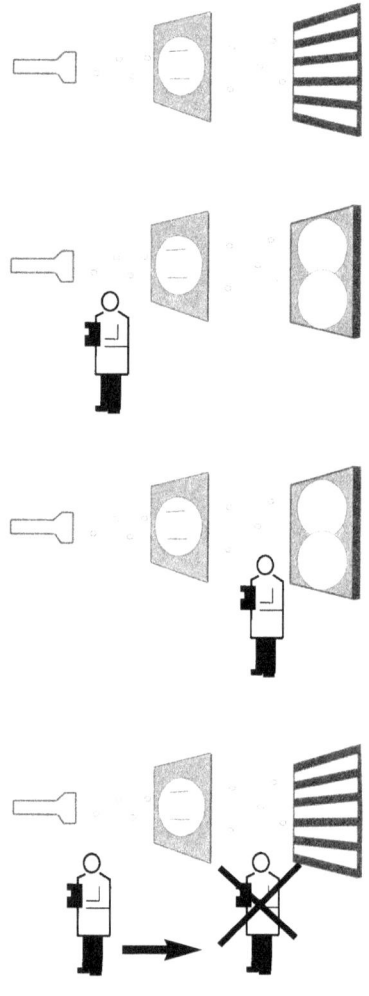

Consciousness creates reality. If the researcher can know which slit a photon passed through, the waves collapse into a pattern that conforms to this potential knowledge.

those conducting the experiment. In the *Newsweek* article reporting on this, Nobel Prize winning physicist Richard Feynman was quoted as saying this is the "central mystery" quantum mechanics, that something as intangible as knowledge—in this case, which slit a photon went through—changes something as concrete as a pattern on a screen.

But how could knowledge change the behavior of particles shot from a gun? Conventional science cannot produce an explanation because the tenet of conventional science previously written about which continues to maintain that the brain produces consciousness, awareness, and thought, which of course means consciousness, awareness, and thought must be confined within a person's skull. Since it would be ludicrous to suggest that thought enclosed inside a person's head could be capable of having an effect on photons shot from a gun. Now, thanks to UVA's research, we know that consciousness is not confined inside the skull. Before UVA researchers went public with their findings, however, many scientists tenaciously clung to the tenet, saying there must be two different sets of laws of physics: a small (subatomic world) set, and a macro world (the one we live in) set. Somewhere between these two worlds, the laws of physics must change. But this assertion does not hold water. First, it doesn't explain why thought contained in someone's head should change things at the subatomic level, or anywhere else. To get around this, scientists came up with the theory that until observed, results of the experiment could take many possible outcomes, and it was the actual observation by a scientist that caused the results of the experiment to collapse into place.

Second, other experiments refute the contention two different laws of physics exist. One such experiment involves large (carbon 60) molecules called "buckyballs," so big they can be seen and therefore are part of the macro world. Research shows they exhibit the same quantum properties as infinitely small particles. Another is an experiment conducted in 2002 that involved crystals. It produced quantum ridges half an inch high—large enough to be measured with a conventional macro world ruler.

What makes more sense is what William of Ockham (c. 1287–1347) is thought to have been the first to say, "The simplest explanation is the best." The researcher's ability to know—his or her consciousness—causes the waves to collapse into particles that form a pattern. In other words, when the researcher can access the knowledge, the zebra pattern does not occur. If he cannot access it, the zebra pattern appears. This was verified by setting up the experiment several ways. In the first, the detectors were in front of the two slits. In the second, researchers placed detectors between the screen and the two slits, i.e., after the photons had passed through the slits. As in the original experiment, knowing about a photon's behavior at the two slits made the zebra pattern vanish. When the detectors were switched off, the zebra stripes returned.

In a third variation, a detector was placed before the slits and a mechanism erased the knowledge after the photon had passed through. The same thing happened. The zebra pattern returned. The result was the same no matter which way the experiment was set up—before the slits, after the slits, or before the slits and then erased. Whether or not the researcher was able to know where each photon hit determined the presence of the zebra pattern, or the lack of it. Versions of the experiment were carried out at the University of Munich and at the University of Maryland. The behavior of the photons, the researchers report, "is changed by how we are going to look at them."

The question is, how? As mentioned above, the answer escaped scientists who refused to think outside the materialist box, but as soon as one is willing to step outside those four walls, the answer is obvious. Consciousness—mind and thought—creates reality. Before it is observed, reality—in this case a multitude of photons fired through slits—exists as non-material potential in the form of waves. Observed, the waves collapse into something more solid—photons that form a pattern. What the researcher conducting the experiment believes ought to be the result, is the result—marks on a screen where the photons hit he or she fired from the photon gun. That fact, dear reader, is the essential truth you need to know: mind creates physical reality.

In other books, I have written about a smoking gun that demonstrates this to be possible. I learned of it in 2008 when I interviewed a tenured professor and chairman of the department of philosophy at the University of Maryland Baltimore County, named Stephen E. Braude, who had recently had a book published by University of Chicago Press (November 15, 2008) called *The Gold Leaf Lady and Other Parapsychological Investigations.* In it, Dr. Braude told the story of Katie, a woman whose mind produced matter—in this case brass: 80% copper and 20% zinc according to analysis, which appeared on her skin in a form that could be compared to the thin foil used to wrap a Hersey's Chocolate Kiss.

Dr. Braude maintains Katie's unconscious mind caused this in an effort to get back at her husband for a humiliating remark. If you are interested, I suggest you read the book and consider the huge implications for quantum physics and the origins of the physical universe that it re-

Who You Are and How You Came to Be

veals. There is also a video on YouTube of Dr. Braude telling the story of Katie the Gold Leaf Lady, which contains footage of brass leaf sprouting from Katie's body.

Stay tuned. In the next chapter we will explore what life on Earth appears to be about.

Chapter Four
What to Do about What You Now Know

Now that you know who you are, what is the best way to make the most of it? I believe it is to take to heart an important point I make in my book *Life After Death Book II, Heaven, Hell & You*. We are all, each one of us, on a hero's adventure. That is why we humans become so enthralled by stories that use it as a basic plot line. A hero or heroine is either compelled or "called" to leave the safety and security of home and venture into the unknown where real or metaphorical dragons and demons must be faced and overcome. *The Wizard of Oz* is a prototypical example. A cave (the witch's castle) must be entered and the broom brought back to the wizard. Supported by unseen or supernatural powers, the hero who pushes forward invariably will succeed, later to return home (the farm in Kansas) more highly evolved than when he or she left and in possession of the elixir, which most often is a new, higher level of understanding. ("There's no place like home.")

This is what life in the physical dimension is about, and why you and I are here. Each of us, and all of us, are on a hero's journey, and once we realize it, we might as well look at life as an adventure that is ours to enjoy to the extent we are able, even during difficult times. To my mind, this makes more sense than thinking of life as full of drudgery and difficulty we must bear and somehow muddle through. Like Dorothy, we have left the comfort of our home in Kansas—in our case, the non-physical world of mind, or "spirit"—and find ourselves in the strange Land of Oz, which in our case is this physical world we have entered via our mother's womb. We might as well accept that during this life we inevitably will face difficulties and struggles and be forced to slay proverbial demons and dragons. In the end, we will return home to the world of spirit wiser and more highly evolved than when we left.

Let me put that another way. Life is one big adventure that will take 70, perhaps 90, or even 100 years. As it unfolds, however, there likely will be a number of smaller hero's adventures—you might think of as side-trip adventures—that will take place along the way. I don't think your overall adventure or the side trips happen by accident. Arthur Schopenhauer (1788-1860), the Nineteenth Century German philosopher, observed that specific events and the meeting of individuals that seemed at the time to have come about by chance, often turn out later to have been essential components in a constant story line. Schopenhauer said that it is as though one dreamer were having a giant dream in which each of the dream's characters has his or her own individual dream. Now we know that your subconscious mind or soul blends with others and the Universal, Infinite Mind, and so it's no wonder the coordination and orchestration takes place. In truth, there's only one dreamer and one mind of which you are a participating part.

It also seems likely to me that your subconscious mind or Soul has certain objectives in this life. It may have a mission it has sent you on, and very likely it has certain learning goals—things you need to experience in order to continue evolving. In several of my books, I write about the character in the movie *Groundhog Day* who keeps experiencing the same day over and over again until he gets it right. The story seems to me an allegory for you and me, meeting the same types of events and having to deal with them until we finally react with love and kindness rather than with contempt. As was the case with Bill Murray's character in the movie, only then will we able to move forward to a new day. For Murray, it was February 3rd.

Look for Unseen Helping Hands

There is an important aspect of the hero's adventure myths I'd like to call to your attention: the unseen hands that typically come to the hero's aid during the darkest part of the story. When all seems lost, the hero that does not give up—but rather, pushes forward with all the effort he or she can muster—will eventually prevail, aided in some way by the hands of fate. These hands of fate are what in Christianity is known as "Grace." This can be seen time and again. Moreover, the Universal Mind will almost always have subsequent events work out so that at least some good will come from even the most tragic events. As the Apostle Paul said, "In all things God works for the good of those who love Him." (Romans 8:28) Coordination happens for the good of all toward the end goal of life in physical form: the evolution of humankind to ever-higher levels of consciousness and self-awareness.

We take on a new role each time we come to Earth, and if we were able to accomplish what we set out to accomplish in our previous life, we will have new goals and new lessons to learn while here. Nevertheless, the basic part of us I call Dharma—a concept that will shortly be explained—likely does not change.

As you recall, Glenn Ford was a music teacher who loved horses, a member of French King Louis XIV's elite horse cavalry, and a cowboy in the American West. Each role was different, but Ford's higher self or Soul appeared to shine through in his affinity for horses. That affinity was an important part of who Ford was, his Dharma—an essential component of his true Self. It seems to me a key to happiness, contentment, and satisfaction in life is to get in touch with that essence. As Joseph Campbell (1904-1987), the American mythologist, writer, lecturer, and professor of mythology and comparative religion, often told his students, "Follow your bliss." That urge, I believe, is something you ought to pay attention to. It is an indication of your Dharma.

When we discover our Dharma and with it the higher part of ourselves, the Creator archetype begins working in our lives because we have establish a connection with the creative source of the universe, the Life force that is the opposite of entropy. Once the connection is made, however, it is important to realize that what your ego-self may wants—a fancy car, a summer home, a promotion at work—and what your true Self wants often are not the same. Your Self with a capital S resides in Spirit and has almost no interest in status, creature comforts, and things of the material world. It's important to honor the desires of the Soul, of course, but I suggest one ought to do so consciously and with an eye toward the potential consequences. If you let your Soul-urges take over completely, you could find yourself with problems that might have been avoided. For example, if you have a wife and family, they may not be happy living in a garret, even though it may not bother you a great deal. I suggest, therefore, that your ego-self and your higher Self form a partnership wherein the needs of each can be satisfied to an extent both practical and possible.

Such a union can be exceptionally rewarding. When someone begins working with his or her Soul rather than resisting its urges as unrealistic or counterproductive, the evolution of that person's Soul will begin to accelerate. Soul growth is the overarching purpose of our lives. Moreover, as we create our lives in cooperation with our Soul, we are contributing to the evolution of humankind toward what Jesus called the Kingdom of Heaven, a state of life on Earth it seems to me Jesus hoped to bring about through his ministry and teachings. It is one in which everyone reacts to and relates with everyone else in a spirit of love and kindness—as Bill Murray's character did on the final Groundhog Day.

Why Bad Things Happen

Like it our not, it is our Souls, not our egos that create our lives. Sometimes it uses a metaphorical two-by-four to get out attention. Your Soul may choose to experience sickness or other loss or suffering as a way to be initiated into deeper wisdom so that it and you might grow. Such choices are an abhorrence to your ego self that typically wants only happiness, prosperity, and higher status. Consequently, you quite naturally feel victimized when this happens, but think about it. If you have been through such suffering, don't you feel you are a better person as a result?

Modern psychological theory stresses how as children we are shaped and formed by our environments, including the culture we are born into, our race, ethnic background, and the socioeconomic status of our family. No doubt these factors weigh heavily in forming the adults that we become. Even so, many students of metaphysics assert that at the deepest Soul level, we choose everything that happens to us including the circumstances of our birth, and in this way our Souls are actually the authors of our destinies. This includes even the most tragic or difficult parts of our lives. Therefore, the more we get in touch with our Souls, the more in touch we become with our true Selves, and the more easily we will navigate through what can be extremely difficult events.

Hard times become even harder when we resist or attempt to turn away in order to avoid the lessons we came to Earth to experience. A major key to a successful life is to realize something I have repeated several times now—that we are an integral part of the Universal Mind. We are in no way separate from it, and that means it is almost always best to—as often has been said—"Go with the flow,"

Who You Are and How You Came to Be

or to employ another cliché, to look for the open door when the we had our heart set on one has closed.

In the upcoming chapter we will how to put what you have just read to use to find your true purpose in life.

Chapter Five
How to Find Your Purpose

    Beyond the lessons you came here to learn, there is another important benefit to getting in touch with your higher self, and that is to determine your purpose in life. It's obvious that practically everyone wants to live a life of purpose. Why else would more than 30 million copies have sold of *The Purpose Driven Life* by Rick Warren?

    How can you find your purpose? An important step is to consciously envision the life you want, a vision so concrete it becomes real to you. It should include putting your Dharma to work to benefit others.

    All self-proclaimed Law-of-Attraction gurus will tell you creating and holding the vision in your mind is paramount. But there is a critical component to manifesting that most do not mention. You should also try to match your vision with the true nature of your Soul and the reality of at least some of the rules of the world you live in. Otherwise, your vision may be merely escapist daydreams. To use an extreme example, it will not help at forty to hold to a vision of someday becoming a professional football or baseball player. The time for that has passed.

    A positive but realistic projection of our future frees us to enjoy life in the present while we make our dreams come true. Visions are most powerful when your family, your friends, or a like-minded group shares them. Napoleon Hill wrote about this phenomenon in his perennially best selling book, *Think, and Grow Rich.*

It's important to be aware, however, that no matter how elevated our consciousness becomes, and however true we are to our higher self, most of us are still confined by our conditioning, by the social constraints of society, and by natural laws. In *Awakening the Heroes Within* (HarperSanFrancisco 1991), Carol S. Pearson wrote, "If we have not taken our journeys and have not developed a strong Ego and connected [it] with our Souls, we are not yet creating consciously." Rather, we are likely to feel, and perhaps we really are the products of our environment and our conditioning.

It may actually be true we are not in control of what is happening to us it we are forced to operate in an oppressive or discriminatory social system or if we live in a dysfunctional family. As Dr. Pearson wrote in the book just mentioned, "Although you might have created the experience of going to jail if you broke a law, that does not mean you created the current reality of the prison system! Much of our lives are created collectively, not individually."

Nevertheless, although many of the circumstances of the life in which you find yourself may be outside your personal control, it makes sense to become conscious of what you can change and what you cannot change and do your best to work within those parameters. Whether or not we are fully the creators of our lives, we are responsible for the degree to which we use whatever power we may have. Each of us came into this life equipped with certain talents and abilities that likely were developed over may previous lifetimes. Taken together, the those abilities and talents come together to form your "Dharma," a Sanskrit word I introduced in the previous chapter, meaning "statute" or "law."

According to a professor at the School of Metaphysics in Windyville, Missouri, Dharma is the law that orders the universe and the essential nature or function of a person or a thing. It is what we as a unique individual have to give or to share with others.

Even though a person may be good at something, that person is not fulfilling his or her Dharma if that person is primarily after acclaim or money. People who are using their Dharma in the most productive ways tend to thoroughly enjoy what they do, and it comes so naturally to them, they do not expect or seek claim because it really isn't important to them.

An acquaintance of mind, Dr. Laurel Clark, who later became the president of the School of Metaphysics, told me during a visit there, "[Dharma] is your soul's urge. When you are responding to your Dharma, you feel at peace. Someday, after you grow old and look back at life, you will regard the time you spent putting your Dharma to work as the golden years. This is because people who are using their Dharma are passionate about what they do, as though it were a flame burning in them. They lose track of time. They're in the flow. And something else. Each person applies his or her Dharma in a way that is unique as though each of us is one piece of a giant jigsaw puzzle and we fit together to make up a whole."

How can you discover your Dharma and thereby your purpose? One way is to think back to the time before you were seven years old and to remember what you loved to do. When you do, it's important to separate what you liked from what was expected of you. Try to remember the different activities you loved and look for a thread that runs through those activities. It also helps to consider the times in your life when you were helping others, having a positive impact, and really felt good about it. Or times when you became lost in an activity and weren't aware of the passing of time. This should deliver important clues.

Remember that as part of the whole of humankind, you are most likely to experience maximum satisfaction from putting your Dharma to work in the service of others rather than for your own self-aggrandizement. Again, this does not mean you need to accept living in a garret as your fate. Your higher self and your ego self need to form a partnership and work out some sort of compromise.

I'm putting emphasis on this because there have been times in my own life when I allowed my Soul-self to run the show, and I can say from firsthand experience, the Soul-self doesn't give a whit about money or creature comforts. Believe me, if there were a few specific turning points in my life I could do over, would take my own advice because losing sleep due to money worries is no fun at all. You see, I discovered my Dharma fifteen years ago and have been employing it ever since—though not always wisely when it comes to finances.

In 2002, a psychic reader at the School of Metaphysics said my Dharma is "Omni-perception." Putting it to work since then has been extremely rewarding in terms of personal satisfaction and the sense of self worth it has created, but unfortunately, it has not delivered the income necessary to live worry-free.

You may be wondering, what in the world is Omni-perception?

To answer, I will describe the psychic reading I had in 2002. The technique employed was virtually the same as that used to produce the more than 14,000 readings given

Who You Are and How You Came to Be

by the man famously known in the twentieth century as, "The Sleeping Prophet," Edgar Cayce (1877-1945). An individual whose psychic ability has been developed is put into a hypnotic trance and questioned. In my case, the one doing the questioning was Dr. Daniel Condron, and his wife, Dr. Barbara Condron, was the person with psychic abilities who gave the reading. This took place in the Moon Valley ranch house on the campus of the School. Tape recorders were cued up. Dr. Dan first spoke Dr. Barbara into a hypnotic trance.

Then he said to her, "You will search the identity of the entity referred to as Stephen Hawley Martin and relate this one's Dharma from the past and past life times in general."

She paused as though waiting for a computer file to boot up, then said in a kind of sing-song monotone:

"This would most easily be described as an omni-perception. There is a very strong urge within this one to interpret that which this one sees. We see that there is a great deal of reliance upon experience but it is from a more distant place rather than an involvement in it [the experience], and we see that this is in an effort to explore and to develop this perception and to answer the urge for it [perception]. We see that there have been many time periods where this one has been in position to be perceptive. There have been instances where this one has been the eyes and the ears of kings. [There was a pause here and a fumbling for words as if she could not believe what she was now seeing or receiving. Then she continued.] This one spent an entire lifetime living in a crow's nest where the entire endeavor was to be able to hone and develop the perception, not only physically but in an otherworldly sense as well.

"There have been many experiences like these that have been building a complete understanding of perception in its omniscient expression. And we see that this one has the ability to see anything from many different points of view. This one has the capacity, therefore, to be able to recognize a whole picture or a whole image where only a fragment is available. This is a very developed and sharpened intuitive sense where this one is capable of experiencing more in a metaphysical sense than what the physical experience itself would allow. Therefore, it is easy for this one to move beyond the limitations of the physical when this one is entrained with the inner mind and with this Dharma. This is all."

Dr. Dan looked at me. Then he picked up a form I'd filled out earlier and turned it over. I'd written a question in the space provided. He turned to Dr. Barbara, "This one would like to know how he can use his Dharma in the present lifetime to serve others."

Who You Are and How You Came to Be

Dr. Barbara said:

"This one is doing so in the ways this one is aware of and the ways this one finds appealing. We see that there is much more that could be done in terms of this one's ability to experience it [his Dharma] in the now rather than linking it to the physical forms of expression. The ability for the perception is keen and will be more keenly developed or directed as this one will become more decisive in terms of the intent of the perception. To this point [in time] much of this one's experience has merely been the receiving of this [perception]. There is recognition that this one is driven to experience many different things with many different people in many different ways and forms and a relishing of this, an appreciation of it that is very deep within this one. There is more, however, that this one can experience with the Dharma by being able to focus the mind upon one point that includes everything. And this [point] is the omnipresence of the perception that this one is capable of in the present time. Therefore, this one would be benefited by beginning to develop inwardly to a greater extent the knowledge of Self to the point of being able to convey this to others. For it is in the conveying of it to others that this one will begin to recognize what this one understands. The interchange is most important for this one, for this is where the greatest opportunity for greater awareness exists. It is in the direct interaction rather than the point of observation. This is all."

Dr. Dan said, "What is the relevance of this ones Dharma to the present lifetime?"

Dr. Barbara answered:

"This one has chosen in the present the conditions whereby there can be the freedom to experience any desire, and many of these have been acted upon, affording this one the availability to experience the omniscience of the perception, and this has brought this one a greater sense of wealth in its true sense. The movement forward would be in the disseminating through interaction of the perception that this one does have. This one has a profound ability as a teacher and a counselor that it would serve this one well to develop."

Dr. Dan asked, "Would it help this one to fulfill his Dharma by being a teacher?"

Dr. Barbara said:

"Yes.

"This is all. . . ."

It had gone by so quickly that I guess I was somewhat in shock.

Omni-perception?

I'd never heard of omni-perception. My thoughts were spinning.

Dr. Dan looked at me. "Do you have any more questions?"

"Uh, yes," I said. "What exactly is meant by omni-perception?"

Dr. Dan turned to Dr. Barbara. "This one asks what is meant by omni-perception?"

She replied:

"Perception is the mental ability to see, to be able to receive what exists. The omniscience in this is to be able to receive all that exists."

Dr. Dan looked at me. "Anything else?"

"Yes. One more thing," I said. "Did I really spend a lifetime in a crow's nest, or was that some kind of metaphor?"

Dr. Dan turned to her. "Was the 'life in the crow's nest' meant literally? A crow's nest on a ship?"

Dr. Barbara said:

"Yes. This one was at sea for almost the entire lifetime."

Dr. Dan said to her, "Is this all?"

Dr. Barbara said, simply, "Yes."

Dr. Dan flipped open the two tape recorders. He handed one of the tapes to me and put the other in a stack.

I returned to my seat on the couch.

Frankly, I was stunned. I'd been aware of my ability to "connect the dots" as I'd called it many times. It was how I got through life. What Dr. Barbara described dovetailed with my speculation that "vision," "discernment" or "comprehension" might be my Dharma. But I hadn't fully realized what this ability truly was, nor had I comprehended the extent to which this, this omni-perception, had been developed. It would take a while to absorb this information and to understand the implications.

One thing was apparent from my reading as well as from the other readings I heard during that weekend at the School. Dharma is not a gift. It is a skill that has been developed over lifetimes.

Imagine spending one in a crow's nest.

And so for the past fifteen years I have been using my Dharma to uncover the truth about life and to communicate what I have discovered by writing and publishing books. As mentioned earlier, it has been quite satisfying, has given my life purpose, but hasn't paid very well.

Before becoming an author and publisher, I made a great deal of money as an advertising executive. But that was an ego-driven job I did not find particularly rewarding. Recently, I have pondered what I could have done to get my ego-self and my soul-self working together to satisfy both our needs, and I believe it would have been to con-

Who You Are and How You Came to Be

tinue the radio show I began in 2007 called "The Truth about Life." It ran on WebTalkRadio.net and within a year was averaging 30,000 listeners per episode.

In 2009, I had an offer from a radio network with 400 affiliate stations to produce and host a live show that would run in a two-hour time slot on Sunday evenings. Realizing how much time and effort it took me to produce a one-hour show each week, and that a two-hour show would require twice the time and effort, I turned down the offer. I also stopped doing the one-hour show on WebTalkRadio and put almost all my effort into my writing. My soul-self enjoyed that activity immensely because it allowed me to analyze information and put my Dharma to work with my writing. I'd started the radio show to promote my books, and it was time-consuming, hard work, the sort of effort that comes along with an ego-driven effort to succeed.

I realize now I should have figured out a way to do the radio show and yet carve out at least an hour or two each day to write because books commonly achieve bestseller status as the result of an author's notoriety. The radio gig could have led to that. One way broadcast personalities get rich is with their books.

Trying to do both would have been a struggle, but it likely would have resulted in a higher income. Suffice it to say that when we give our Soul full rein to create our lives, we often do so by consciously repressing the ego's concerns about money and practical matters. As someone who has spent long stretches fulfilling ego urges, and other long stretches fulfilling Soul urges, my advice is to honor and listen to the wisdom of both and figure out a way to satisfy each to the extent possible.

Chapter Six
The Return of a Very Old Worldview

More than twenty years ago, I read a book by W. E. Butler (1898-1978) whose description of the heroes' journey we humans collectively are on has remained with me. He said the evolution of humankind can be compared to a great crowd of people making its way along a road that winds up a hill, plodding along as a flock of sheep might, kicking up dust but moving slowly, stopping now and then, scrapping and biting one another, now and then getting panicky and shifting one way or the other, often hardly moving ahead at all. That description resonated with me. Hopefully, we now have reached a point in terms of knowledge and understanding when we will take a giant leap forward, which is what this final chapter is about.

If you have read my book, *Life After Death, Powerful Evidence You Will Never Die,* please accept my apology because what comes next is taken from that book, much of it verbatim. It may be useful, however, for you to read it again because it seems to me that if my theories are correct, and of course I think they are, once the theories are accepted—which I realize may be quite some time in the future—the human race will have come full circle.

This is the case with all heroes' journeys. Ultimately, the hero returns home, just has the human race will return home, but on a higher plane of understanding than when the hero left. So you might say, the hero does not journey out and back in a circle, but rather, in a spiral that takes him or her to a new and higher level.

There was a time, anthropologists tell us, when humans felt at one with nature, a feeling I believe may come again. Called pantheism, humans long ago knew they were an integral part of the Universe. The Divine displayed itself in many forms and was present in all things. But as humans grew more self aware, they began to feel separate and apart. As stated in Chapter One, the allegory of Adam and Eve recalls the time when humans parted company with the view that they could or should commune daily with the Divine. They cut the cord by exercising free will.

No longer seeing God in themselves and in others, we humans conjured up gods outside ourselves. In ancient Greece, for example, many gods representing various human qualities were thought to exist. The worldview that evolved in those ancient times had man in the middle between two worlds—a place the Chinese referred to as the Middle Kingdom. The gods lived above the clouds of Mt. Olympus, although they did come to earth now and then, mostly to cause problems for humans.

Who You Are and How You Came to Be

Below the Middle Kingdom—what caused it to be in the middle—was the underworld, home of the dead, where Hades was in charge and the three-headed dog Cerberus guarded a gate one came to after crossing the River Styx.

Different cultures had different takes on this three-layered universe. Then as now, ideas about God and gods differed depending on the group one belonged to. The Egyptians had Bal. The Jews had the god of Abraham. The Romans and the Greeks had a pantheon full.

The Idea of One God Created a New Worldview

Then came Jesus of Nazareth and the idea emerged that only one God ruled over creation—although He did have angels and eventually saints who took up some of the positions left vacant by departing Roman and Greek gods.

In 1994 Karen Armstrong published a book, *A History of God,* which chronicled history of the emergence of the concept of one God.

Because of this idea, the worldview changed. God and angels replaced the pantheon of gods above the clouds. A fallen angel, Satan, replaced Hades. The place below the ground became hell, rather than the underworld—where evildoers went. The good folk would be raised at the end of time on Judgment Day.

This view held sway for more than a thousand years but was destined to change again because of a new scientific discovery by Christopher Columbus (1451-1506).

Columbus lived on high ground overlooking a Mediterranean harbor. I have visited the ruin of what is said to be the house where he grew up. In that part of the world there is almost no humidity and the air is very clear. If Columbus had good eyes, he would not have needed a spyglass to see ships climb over the horizon as they approached the harbor. I've witnessed this myself. Columbus could see for himself the world was round, and he must have decided to prove it by sailing west to get to the spice islands of the East Indies.

Columbus never realized it, but he didn't actually get there. Nevertheless, some of Ferdinand Magellan's (1480-1521) crew did, and beyond. Of the 237 men who set out on five ships in 1519, eighteen actually completed the circumnavigation of the globe and returned to Spain in 1522.

The newly realized fact that the world was round forced the then commonly held worldview to change. Nevertheless, since people, and more important, Church leaders believed that God had created it, the earth remained at the center of the universe. Now heaven, the dwelling place of God, was seen as being somewhere above the stars. Hell remained beneath the ground, down where it was hot, the place from which molten lava spewed when volcanoes erupted.

The Worldview Gets an Update

It wasn't long before this worldview had to be updated. A fellow named Nicolaus Copernicus (1473–1543) determined the sun was at the center of the solar system. But the Church—the authority back then as science is today—pretty much ignored this concept because it did not go along with accepted canon.

Then, a century later, along came Galileo Galilei (1564–1642), a man who would not leave well enough alone. Galileo—among other things an astronomer—championed Copernicus's assertion as proven fact. As a result, Galileo started having to watch his back. This was heresy. At that time people were being burned at the stake for less. Indeed, the leaders of the Church told Galileo he'd better recant, and he did. As a result, Galileo got off relatively easily, spending the final years of his life under house arrest on orders of the Inquisition.

But even the Church could not keep word from getting out. Gradually, the accepted views of the day began to change.

Who You Are and How You Came to Be

A Tiny World Comes into View

In 1675, a Dutchman named Antoni van Leeuwenhoek (1632-1723)—an amateur lens grinder and microscope builder—saw for the first time tiny organisms he called "animalcules" living in stagnant water. He also spotted them in scum collected from his teeth. Leeuwenhock didn't know or even speculate that "animalcules" might cause disease. It took until the nineteenth century for that revelation to dawn. At the time, the idea creatures so small they were invisible to the naked eye entered the body to make a person sick and sometimes die would have seemed totally absurd, just as the idea the brain does not create consciousness may seem to some today. It was thought that demons and the devil caused disease, or that God did it to punish sinners. In 1692 in Salem, Massachusetts, eighteen were hanged and one was crushed to death because they were thought to be witches in league with Satan. No wonder after that, and down until today, the idea of Satan and demons and witchcraft was thought to be pure superstition. To believe in such things was to invite witch-hunts and mass hysteria, and nobody wanted that.

Who You Are and How You Came to Be

The Age of Reason Dawns

Even so, a new day was dawning, a period alternately referred to as "The Age of Enlightenment" and "The Age of Reason." As was discussed in Chapter One, the English philosopher, Thomas Hobbes (1588-1679), had argued that aside from God—the "first cause" who created the material world—nothing existed that is not of the material world. The logic he used was simple. How could it if God created everything?

This view was ultimately to lead to the great clockmaker theory, the idea that God created the universe, wound it up, let it go, and was no longer involved in its operation. Natural laws also had been created that kept going what had been set in motion. Called Deism, many founding fathers, including my personal hero, Thomas Jefferson, subscribed to this.

Hobbes had a big impact on the Age of Enlightenment, which was to pick up steam in the eighteenth century. But the big kahuna was Sir Isaac Newton (1643 – 1727), an English physicist, mathematician, astronomer, natural philosopher, alchemist, and theologian. Certainly one of the most influential men of all time, his *Philosophiæ Naturalis Principia Mathematica,* published in 1687, is considered to be the groundwork for most of classical mechanics. Newton described universal gravitation and the three laws of motion that dominated the scientific view of the physical universe until the advent of quantum mechanics. It seems safe to say Thomas Hobbes's materialistic view of reality coupled with Newton's mechanistic view is the bedrock of scientific thinking today, except among quantum physicists and metaphysicians.

The prevailing worldview that emerged from the Age of Reason was that the universe might be compared to a giant machine. The Sun was at the center of the solar system. The Earth and planets revolved around it. Nothing existed but the material world. What was thought of in the seventeenth century and earlier as the invisible world of spirit did not exist. Everything that happened had a logical cause. Natural laws governed everything.

Darwin's Theory Takes Hold

In 1859 an Englishman, Charles Darwin (1809-1882), published *On the Origin of Species,* a seminal work in scientific literature and a landmark work in evolutionary biology. Its full title, *On the Origin of Species by Means of Natural Selection, or the Preservation of Favoured Races in the Struggle for Life,* uses the term "races" to mean biological varieties. Darwin's book introduced the theory that populations evolve over the course of generations through a process of natural selection. It presented a body of evidence indicating the diversity of life arose through a branching pattern of evolution and common descent. In other words, God had not created the variety of life on the planet, nor had He created humans. All this had happened through a natural—what might be seen as mechanical—process. This became accepted as fact among the educated classes.

But astute scientists then and now realized something important is missing from Darwin's theory. It cannot be reconciled with the Second Law of Thermodynamics, also called the Law of Entropy—the fact that in a closed system things tend to break down and fall apart, rather than get better.

How then could life get more complex by accident? What caused an eye, a kidney, a heart, ears, and all those complex systems to develop? I believe as stated in this book it had to do with the Life Force—the underlying intelligence, subjective mind, opposite of entropy that also manifests as Grace. But only a few thinkers, notably Thomas Troward (1847-1916), considered such possibilities back then. Most ignored his theory and overlooked the flaw. Many still do today.

Darwin's theories reinforced the rationalist idea that the so called supernatural was a figment of human imagination and—not wanting to be burned at the stake—most scientists probably were happy to keep it safely buried. Life and its diversity were results of a natural process that also conveniently justified nineteenth century Imperialism. After all, "Natural Selection" was also known as "Survival of the Fittest," and what could be more fitting than white men with guns bringing civilization to "primitive" people of color—red, black, brown, and yellow—all around the world? Intelligence and mind had evolved as life had evolved and had reached its pinnacle in humans, especially the white ones. Consciousness and intelligence were produced by an organ, the brain, which had come about through evolution.

A Wedge Is Driven Between Science and Religion

With this worldview, a wedge was inserted and hammered in between science, religion and any possibility of things so called supernatural. Hobbes had said nothing existed but the physical. If this were so, where could God possibly reside? What about the heavenly hosts? Thought was contained within the skull so what possible good could prayer do?

A line was drawn. Educated men and women could not believe in God and prayer or angels or ghosts and demons, which were seen as figments of imagination, ignorance, and superstition. Many may have had a yearning for

Who You Are and How You Came to Be

God—as humans seem to for things spiritual—but could not rationalize His existence. All were forced to choose between religion and science, though many attempted to straddle the line—as many still do today.

Now, in the first part of the twenty-first century, this worldview continues to be the only socially acceptable one in some circles. But there are signs it is beginning to crumble. Hundreds of thousands, perhaps millions, have shifted to a new worldview based on a new branch of science called quantum mechanics and the findings of scientific research that do not fit the materialist-reductionist mold. My hope is that this book will play an important role in helping to put Scientific Materialism on the ash heap of history where it belongs along with other once strongly-held yet bogus beliefs such as the earth is flat and the sun revolves around it.

Let's look at some of the pioneers who have not been afraid to speak out, as well as their ideas and discoveries that conflict with the prevailing nineteenth and twentieth century worldview. The following does not in any way represent an exhaustive list. My apologies to anyone who feels left out, and to anyone who thinks I have overlooked a key figure.

Who You Are and How You Came to Be

Matter = Energy and Universal Mind

In 1905, Albert Einstein (1879-1955), a German-born theoretical physicist, published a paper proving that light behaves both as a wave and as particles. This, as well as Einstein's famous formula, $E = MC^2$, indicates reality and matter are not what they seem. Matter or mass as it is referred to in this formula is equivalent to energy and vice versa.

In 1912 Swiss psychiatrist Carl Jung (1875-1961) published *Wandlungen und Symbole der Libido* (known in English as *The Psychology of the Unconscious*) that postulated a collective unconscious, sometimes known as collective subconscious. According to Jung there is an unconscious mind shared by a society, a people, or all humanity, that is the product of ancestral experience and contains such concepts as the classic archetypes, science, religion, and morality.

Quantum physicists came along who expanded on Einstein's work. Niels Henrik David Bohr, a Danish physicist, made fundamental contributions to understanding atomic structure and quantum mechanics, for which he received the Nobel Prize in Physics in 1922. He is quoted as having said, "Everything we call real is made of things that cannot be regarded as real."

Who You Are and How You Came to Be

Nothing is really solid. Everything is energy—vibrations.

ESP and Psycho Kinesis Are Proven Real

As we know from our earlier discussion, in the early 1930s a man named J. B. (Joseph Banks) Rhine moved from Harvard University to Duke to set up a parapsychology laboratory. Rhine not only founded the parapsychology lab at Duke, he also founded the Journal of Parapsychology and the Foundation for Research on the Nature of Man. His double blind studies conducted largely between 1930 and 1960 established that ESP exists and is real. Not mentioned in our earlier discussion, they also showed psycho kinesis—mind over matter—is real as well, at least to a small degree.

His findings were either scoffed at or ignored by the scientific community then as they continue to be today.

Zen Is Introduced to the West

In 1953, Eugen Herrigel (1884-1955), a German philosopher who taught philosophy at Tohoku Imperial University in Sendai, Japan, from 1924-1929 published the book, *Zen and the Art of Archery*. This introduced Zen Buddhism to the West and the concept that "All Is One," i.e., everything is connected rather than made up of separate parts.

How else could Zen masters shoot arrows while blindfolded and consistently hit the bull's-eyes of targets many yards away?

In 1966 a British philosopher named Alan Watts (1915-1973) published a book called *The Book: On the Taboo Against Knowing Who You Are* that went into detail about Buddhist thought. Known as an interpreter of Asian philosophies for a Western audience, Watts wrote more than 25 books and numerous articles on subjects such as personal identity, the true nature of reality, higher consciousness and the meaning of life. His writings and ideas fueled a new movement that came to be known as New Age.

Plants Tune into Thoughts

As discussed, a polygraph expert named Cleve Backster (1924-2013) began research in 1966 that demonstrated living plants tune into the thoughts and intentions of humans as well as other aspects of their environments, thus indicating some sort of hidden mental connection between living things. His findings were ridiculed, but have since been confirmed by other researchers.

In 1978 a young man with a B.A., M.A., and Ph.D. from the University of Virginia and an M.D. from Georgia Medical School named Raymond Moody (born 1944) published a book called *Life After Life,* in which he detailed the experiences of people who had been clinically dead and resuscitated.

The Phenomenon of Grace Is Publicized

Also in 1978, a psychiatrist named M. Scott Peck (1936-2005) published a book that became a huge bestseller called, *The Road Less Travelled: A New Psychology Of Love, Traditional Values And Spiritual Growth*. Among other things, Peck's book dealt with the phenomenon of grace, which we covered in the last chapter. He said grace was both common and to a certain extent, predictable. He also wrote that, "grace will remain unexplainable within the conceptual framework of conventional science and 'natural law' as we understand it."

Grace is the unseen force that brings the best possible results out of unfortunate events and circumstances, i.e., "every cloud has a silver lining." In Peck's own words, "There is a force, the mechanism of which we do not fully understand, that seems to operate routinely in most people to protect and encourage their physical health even under the most adverse conditions." His book gives specific examples.

Although I repeat myself, it seems to me, Grace is the Life Force at work.

Who You Are and How You Came to Be

Quantum Physics Is Introduced to the Masses

In 1979, Gary Zukav, a former Green Beret during the war in Vietnam, published a book called *The Dancing Wu Li Masters: An Overview of the New Physics*. Targeted for laymen, it explained the basics of quantum physics in everyday language, i.e., without the use of complicated mathematics. Zukav concluded that "the philosophical implication of quantum mechanics is that all of the things in our universe (including us) that appear to exist independently are actually parts of one all-encompassing organic pattern, and that no parts of that pattern are ever really separate from it or from each other."

Also in 1979, James Lovelock published a book called *Gaia: A New Look at Life on Earth* that explained his idea that life on earth functions as a single organism. In contrast to the conventional belief that living matter is passive in the face of threats to its existence, the book explored the hypothesis that the earth's living matter—air, ocean, and land surfaces—forms a complex system that has the capacity to keep the Earth a fit place for life. Since Gaia was first published, many of Jim Lovelock's predictions have come true.

The Spiritual Dimension Is Explored

In the mid 1980s a television series appeared on PBS called *The Power of Myth,* featuring author and Sarah Lawrence College Comparative Religion Professor, Joseph Campbell (1904-1987). These programs made an impact on a significant segment of the public and opened their eyes to the possibility of the existence of what might be termed "a spiritual dimension." This can be summed up using Campbell's own words, "Anyone who has had an experience of mystery knows there is a dimension of the universe that is not available to his senses."

Scientific Studies Demonstrate the Efficacy of Prayer

In July, 1988, Dr. Randolph Byrd, a cardiologist, published an article in the *Southern Medical Journal* about the effects of prayer on cardiac patients. Over a ten-month period, he used a computer to assign 393 patients admitted to the coronary care unit at San Francisco General Hospital either to a group that was prayed for by home prayer groups (192 patients), or to a group that was not prayed for (201). A double blind test, neither the patients, doctors, nor the nurses knew which group a patient was in.

The patients who were remembered in prayer had remarkably, and a statistically significant number of better experiences and outcomes than those who were not prayed for. Also, fewer prayed-for patients died, although the difference between groups in this case was not large enough to be considered statistically significant.

In 1994 Rupert Sheldrake, a British biochemist, published a book called *A New Science of Life,* which the editors of the British journal, *Nature,* called, "the best candidate for burning there has been for many years." In a nutshell, Sheldrake's theory is that living things, humans included, are formed by what he called morphogenetic fields in conjunction with genes. In essence, Sheldrake argues that genes determine the proteins or building blocks, and morphogenetic fields the blueprints of the many different organisms and life forms. Each species has a field, as does each individual human being, and these fields continue and evolve from generation to generation, thus providing a scientific explanation for reincarnation as well as evolution. I go into this in much more detail in other books I have written.

Researchers' Knowledge Determines Outcomes

In 1995, Raymond Chiao, a Hong Cong native and quantum physicist then teaching at the University of Cal-

Who You Are and How You Came to Be

ifornia at Berkeley, published a paper about a series of experiments. As stated in Chapter Three, the paper reported upon in a July 1995 issue of *Newsweek* magazine, said that what researchers knew or did not know about certain aspects of each experiment had a predictable determination on their outcomes. In other words, what was in the researchers' minds—thought—apparently determined the result. In the *Newsweek* article reporting on this, Nobel Prize winning physicist Richard Feynman (1918-1988) was reported as having said this is the "central mystery" of quantum mechanics, that something as intangible as knowledge—in this case, which slit a photon went through—changes something as concrete as a pattern on a screen.

Prayer Adds Fuel to the Life Force

Also in 2001, a study published in the September issue of the *Journal of Reproductive Medicine* showed that prayer was able to double the success rate of in vitro fertilization procedures that lead to pregnancy. The findings revealed that a group of women who had people praying for them had a 50 percent pregnancy rate compared to a 26 percent rate in the group of women who did not have anyone praying for them. In the study—led by Rogerio Lobo, chairman of obstetrics and gynecology at Columbia University's

College of Physicians & Surgeons—none of the women undergoing the IVF procedures knew about the prayers on their behalf. Nor did their doctors. In fact, the 199 women were in Cha General Hospital in Seoul, Korea, thousands of miles from those praying for them in the U.S., Canada and Australia. This collaborates with other studies and quantum physics theory that distance is not a factor at the subatomic level of mind.

Research Tells How Best to Pray

An organization exists that has as its purpose the study of what prayer techniques produce the best results. It's called Spindrift and was founded by Christian Science practitioners who have been at this since 1975.

The first question Spindrift researchers sought to answer is, does prayer work? The answer, as we already know, is yes. In one test, rye seeds were split into groupings of equal numbers and placed in a shallow container on a soil-like substance called vermiculite. (For city dwellers, this is commonly used by gardeners.) A string was drawn across the middle to indicate that the seeds were divided into side A and side B. Side A was prayed for. Side B was not. A statistically greater number of rye shoots emerged from side A than from side B.

Who You Are and How You Came to Be

Variations of this experiment were devised and conducted, but not until this one was repeated by many different Christian Science prayer practitioners all of whom got consistent results.

Next, salt was added to the water supply. Different batches of rye seeds received doses of salt ranging from one teaspoon per eight cups of water to four teaspoons per eight cups. Doses were stepped up in increments of one-half teaspoon per batch.

A total of 2.3 percent more seeds sprouted on the prayed-for side of the first batch—one teaspoon per half-gallon of water—than on the not-prayed-for side—800 "prayed-for" seeds sprouted out of 2,000, versus 778 sprouts out of 2000 in the not-prayed-for side. As the dosage of salt was increased, the total number of seeds sprouting decreased, but the proportion of seeds that sprouted on the prayed-for sides increased, compared to the not-prayed-for sides, as the amount of the salt—stress—increased. In the 1.5 teaspoon batch, the increase was 3.3 percent. In the 2.0 teaspoon batch, 13.8 percent. In the 2.5 batch, 16.5 percent. In the 3.0, 30.8 percent. Five times as many prayed-for seeds in the 3.5 batch sprouted—although the total number which sprouted was small as can be seen from the chart below. Finally, no seeds sprouted in the 4.0 teaspoon per eight cup batch.

What this says is what people lying in a ditch with bombs going off around them have always known: the more dire the situation, the more helpful prayer will be. Up to a point. There comes a time when things are so bad nothing helps.

Studies similar to this have been and are being carried out by a consortium of scientists put together by Lynne McTaggart, author of the book published in 2002, *THE FIELD: The Quest for the Secret Force of the Universe,* and her 2008 release, *The INTENTION EXPERIMENT: Using Your Thoughts to Change Your Life and the World.* When she was on my show in early 2008, she described some of these experiments and the terrific success she and her colleagues have had. She said several of these studies were already being prepared for publication.

Mediums Relate Information about the Dead

Also in 2008, Julie Beischel, Ph.D., published a paper in *The Journal of Parapsychology* in which she concluded, " . . . certain mediums can report accurate and specific information about the deceased loved ones (termed "discarnates") of living people (termed "sitters") even without any prior knowledge about the sitters or the discarnates and in the complete absence of any sensory sitter feedback.

Moreover, the information reported by these mediums cannot be explained as a result of fraud or 'cold reading' (a set of techniques in which visual and auditory cues from the sitter are used to fabricate 'accurate' readings) on the part of the mediums or rather bias on the part of the sitters."

Toward a New, Higher Level of Understanding

In the past, a single discovery could create a new worldview—that the earth was round, or that the sun was the center of the solar system. Then Newton's laws followed by the *Origin of Species* did the trick. It seems to me so many discoveries have occurred since then that more than enough are on the table now to create a new one.

What's holding us back? Those with a vested interest in maintaining the status quo. People who do not want to look stupid. Ignorance on the part of people too caught up in the information overload to see the forest for the trees. People with dogmatic religious beliefs. The truth is, a new worldview is already held in part by many, and in full by a small percentage of the population in the West today.

Ironically, as was stated at the outset of this chapter, the new worldview is not fundamentally different from that which existed before self-awareness caused humans to feel separated from the rest of nature. It is that mind—the intelligent medium of thought—is the ground of being of all that is, and that we and everything in the universe are not only connected to it and to each other, we are each aspects of it. We are at one with nature; part of One Mind. In our new understanding that the Divine animates us all, we have come full circle and have arrived on higher ground in our understanding of who we are and how we came to be.

About the Author

Stephen Hawley Martin is a professional writer and ghostwriter, the winner of a number of national awards for his work, and the Editor and Publisher of The Oaklea Press.

You can learn more about Steve and his books on his website: *www.shmartin.com,* and on YouTube by putting his full name in the search bar.

You may also wish to visit *www.OakleaPress.com* to find out how you can work with Steve to bring your book to reality.

NOTES

www.ingramcontent.com/pod-product-compliance
Lightning Source LLC
Chambersburg PA
CBHW070316230526
45470CB00002B/900